Twelve Sonnets:

A Defense of Spirit

Kevin Farnham

Twelve Sonnets: A Defense of Spirit

Cover Design: Dale Farnham
Cover Painting: "The Storm" by Pierre August Cot, 1880
Painting Image Source: WikiCommons

Editor: Dale Farnham

ISBN: 978-0-9778833-6-3

Published by Everush Books

This book is available at EverushBooks.com and at online booksellers. The book may also be ordered through retail book stores using the ISBN number.

For Dale

Incipit vita nova—

"Here begins new life."

With other ministrations thou, O nature!
Healest thy wandering and distempered child:
Thou pourest on him thy soft influences,
Thy sunny hues, fair forms, and breathing sweets,
Thy melodies of woods, and winds, and waters,
Till he relent, and can no more endure
To be a jarring and a dissonant thing,
Amid this general dance and minstrelsy.

—Samuel Taylor Coleridge
"The Dungeon"

Acknowledgement

"Ageless Nymph" (Sonnet 2) originally appeared in "The Lyric" Volume 103, Number 1 (Winter 2023).

Contents

Context

Writing in this Third Millennium AD, where we've supplanted Nietzsche's "God is dead" with an expanding vision of biology and psychology as being machines, I find it necessary to begin this work with a justification of art as being meaningful and relevant today.

In my early twenties, I was in a very depressed state intellectually, believing that God indeed does not exist, and believing that the laws of Physics, particularly the Second Law of Thermodynamics, prove that life is meaningless: the Universe is in a steady and irrevocable regression to a state with zero movement at a temperature just above absolute zero. If this hopeless situation is truly the case—if existence is meaningless—then why endure the often excruciating suffering of life? Indeed, the logically valid protest against participation in such a living death is suicide. The problem with such an action, though, is the enormous added suffering it would bring to others: it would be a selfish betrayal of the love they'd selflessly showered over me.

1

My viewpoint was forever changed when I read Owen Barfield's *Saving the Appearances: A Study in Idolatry.* Barfield points out that science, as a discipline, has a very specific and limited role: to identify those aspects of nature that can be represented as mechanism. To "save" an appearance, in the Middle Ages, was to come up with a mathematical formula that models a particular observed feature of nature (for example, the movements of the planets in the night sky). This was considered to be the sole and exclusive role of science, within the wider realm of human inquiry and knowledge.

Barfield calls it a mistake, a form of idolatry, when we consider science as extending beyond these limited bounds, when we consider scientific models, the psychological machines science invents to depict the processes of nature, to be the sum total and end all of human knowledge, and reality. To say nature is a machine and nothing more is akin to molding a golden calf and saying "This is God." It's idolatry.

It appears that we haven't advanced all that far with respect to idolatry over the past many millennia. In order to wake up from our current situation, this false nightmare with which we today weigh down our psyches, Barfield argues that we need to "smash" this idol— the idea that the Universe (including our own being) is a machine and nothing else—so that we may re-learn to appreciate the full reality of what Nature is, the reality of what our own being is, appreciate who we are both interiorly and within the fullness of Nature.

It doesn't appear that Barfield's view has gained many adherents in the past more than half century since he wrote *Saving the Appearances*. If anything, we've drifted farther along the path of viewing ever more aspects of nature (for example, our bodies and personalities) as mechanisms and probably nothing more.

Hence (in part), the present book. Here, I attempt to illustrate (if not prove) that spirit is real (both in nature and within ourselves), testifying to the smashing of the idols that once dominated my own thought (which continue today to dominate the thinking of a great many intelligent people), and laying a philosophical groundwork and context to justify the future sonnets, and all true art.

The sonnets in this book were written starting more than a decade ago, as the beginning of a project I called *The Autumn Sonnets*. The sonnets would document the autumn years of my own life, centering on the daily communion I live with my wife, and also include reflection on the waning vitality of Western Civilization (compared with what has come before).

When I decided these initial sonnets should be a

stand-alone book, I looked to Dante Alighieri's *La Vita Nuova* as a structural model. Like Dante's early work, *Twelve Sonnets: A Defense of Spirit* is a prosimetrum, a collection of poems alternating with counterpoint prose that connects the poems and amplifies their content.

Prosimetra have a long tradition in both Eastern and Western literature. Western examples include *Satyricon* by Petronius from the First Century AD; *On the Consolation of Philosophy* by Boethius around 524 AD; *Cosmographia* by Bernard Silvestris (1174); Dante's *La Vita Nuova* (1295); *Ascent of Mount Carmel* by Saint John of the Cross (circa 1579); *The Countess of Pembroke's Arcadia* by Philip Sidney (1593); *Spring and All* by William Carlos Williams (1923); and Vladimir Nabokov's 1962 novel *Pale Fire*, in which the chief "characters" are the prose and verse components of a prosimetrum.

The poems in this volume ask much from the reader. You are expected to know about, or at least be willing to learn something about, subjects ranging from Greek mythology to the histories of literature and science, philosophy, world religions, and modern physics. For this reason the poems are supplemented with (sometimes copious) notes.

The present book is precursor to, and also the beginning of, *The Autumn Sonnets*. This small prosimetrum sets the ground for the larger work, which will be almost exclusively poetry. The prose discussion here may assist the reader in understanding the poetic method that will be utilized and more fully actualized in the complete work.

Presupposing that the ground work of justifying art is indeed laid out in this book (i.e., it is proved here that art is a valid form of inquiry which can be a repository of knowledge and exhibit truth), and trusting that the reader is willing to undertake some research beyond the poems in order to fully comprehend the references and allusions, what they mean and signify, why they're in the poems, we can begin.

Sonnet 1:
Life History

My name was Merlin: him you chose to woo.
I spoke in visions ("Time's an ushing well"),
and held both gods and science in my spell,
and in my mind would all the world construe.
The books! O, shining books I'd write for you:
of courtship, wedding, children, nature's peace—
taut words creating light without surcease,
that every moment soul's deep core renew!
Now I am old—indeed were dead, had not
our alchemists of Physic wrought a cure.
This remnant dust can't conjure Camelot;
still, may one word (our hearts' embrace) endure.
Henceforth, Love, autumn sonnets I'll compose:
In these our life, our time, this world enclose.

Notes

I had originally planned a literary career in the mold of Joyce or Dante. But the needs of children interrupted composition for decades, though my studies in preparation for the writing continued.

I was still working hard in the world, my engineering career, still planning to ultimately accomplish the long-planned oeuvre (thanks to today's much increased human longevity) when, of a sudden, I had a heart attack—having actually written almost nothing despite having spent decades in studious preparation.

Thanks to modern medicine, I was revived. But it suddenly struck me that the remaining available time might be too minimal to provide for completion of the original oeuvre. I decided that, to have a chance to convey at least some portion of the requisite message, my writing would have to be brief and concise.

After long consideration, looking back across the entirety of Western literature, noting in particular what others had managed to say in very few words, I found something that appeared still possible: focus especially, succinctly, and intimately on life today with my wife, my beautiful and wise Baucis, in these our autumn years (which also appear to be a twilight in Western Civilization).

Merlin is a prophet, wizard, bard in legends of King Arthur; see, for example, Geoffrey of Monmouth, *Historia Regum Britanniae* (History of the Kings of Britain), circa 1136.

words creating light refers to my concept of the poet's mission: to transmit what is seen interiorly to those who have not (yet) seen that way. The best way to do this is to translate the meaning of one's interior visions into a story, an image, that is comprehensible by others.

Alchemy is the human-willed changing of one material substance into another—in this case, changing physical illness into physical health.

Physic was originally the study of natural things, later the study in particular of healing. From the Greek φυσικος. An early English example of the word's use is in John Gower's *Confessio Amantis* (The Lover's Confession), circa 1392: "Ther is phisique for the seke" (Middle English "seke" in this case equates to modern "sick").

Camelot is King Arthur's castle and court, a celestial capital of wisdom and justice. See, for example, Chrétien de Troyes, *Lancelot, the Knight of the Cart*, circa 1181.

Sonnet 2:
Ageless Nymph

The ageless nymph knows nought but to educe
life's art: perform her garden ministry,
mindfully cipher Earth's geometry—
counting, pruning, making the world profuse.
I fear to stir: what turn might that induce?
My modern glance assaults her secrecy:
she starts, pales, trembles, darts away from me,
sensing I once applauded her abuse.
Trampled by zealous gods, she fears the light:
eternal being, that one true Faith distorted;
then, further, Science's piercing lens contorted—
excising Nature's spirit from our sight.
Sad form: cross centuries deep you've watched and pined
as logic's triumph warped the Western mind.

Notes

The evolution of human consciousness is illustrated from the point of view of living Nature, as personified by a delicate nymph.

Contemplating nature today, a student of poetry who is also a scientist notices a near absence of perceptible spirit. In Owen Barfield's phrasing, science over recent centuries has "scoured" the phenomena we sense of spirit, leaving behind a brittle skeletal logical structure.

Scientific logic says "There's nothing there but particles"; yet poetic vision says "There *must* be spirit there— else all nature poetry from the past was mere psychosis!"

Nature hasn't changed much across recorded human history. What's changed, evolved, is human conception about what Nature is.

Today we struggle to experience the spirit that's in Nature because the universe-as-mechanism logic in our minds influences the perceptual images that are formed within us. It's as though our modern scientific thinking is a lens that, like Sylvia Plath's bell jar, distorts our vision of the true Nature that surrounds us, filters away its life.

Sonnet 3:
Stately Elders

Sad Hero welcomed death: the winter gale
had swallowed her Leander. Thisbe yearned,
till hope was by a lioness overturned.
Troilus, sad Troilus: hear his stricken wail!
Tibullus found pure love could not prevail;
they fled from Wyatt; Sidney's furnace burned;
Daniel, Catullus, Drayton, Yeats were spurned.
Who but a fool would love? To what avail?

Two stately elders grace our spare expanse;
their branchlets, twined, caress and sway, impelled
by summer's aether: splendrous pantheon!
Just so, our naked souls embrace and dance
in light that breathes within: our spirits meld!
Love, you're my Baucis; I'm your Philemon.

Notes

Ancient myth is filled with examples of lovers whose love led them to ruin. Likewise, poets across the ages documented their suffering when their desire for union was stifled, as they were repeatedly and forever rebuffed by the one they felt they needed to be with to the exclusion of everyone else. Given all this love-induced suffering, would not a sensible logic be to simply avoid romantic love entirely?

But this is not the case for us, my love.

Hero and Leander were young lovers. The swirling gulf of the Hellespont (today's Dardenalles) separated them. In her tower, each summer night, Hero held up a light, to guide Leander as he swam across the strait to be with her. Winter winds make the Hellespont's waters much rougher; they are much colder, too. One winter night the lovers' burgeoning impatience eclipsed prudence. Leander thought he saw Hero's light. He stripped and began to swim. The light disappeared, perhaps extinguished by wind. Leander drowned, and his body was washed toward shore near Hero's tower. In the dawning light, Hero saw her Leander, dove into the waters, clasped him, and drowned. Later, they were found dead on the shore, wrapped in one another's arms.

Thisbe and Pyramus lived in houses separated by a

wall. Their parents would not have approved their being in a love relationship. They found a small crack in the wall and conversed and loved through the wall day and night. One day they plotted to meet one another that night in a grove near a tomb. Thisbe arrived to find a lioness that had just finished eating its bloody prey. She dropped her scarf while running to hide. The lioness noticed the scarf and tore it to shreds, leaving it blood-drenched on the ground. Pyramus arrived, recognized the scarf, and, assuming Thisbe had been killed, stabbed himself with his dagger, his blood spurting and splashing the leaves and berries of a mulberry tree above. Thisbe returned, noticed the strangely altered berries, then saw Pyramus on the ground, holding the bloody scarf, dying. She cried out to him, watched him die, then took the dagger, pointed it upward, and pressed her body onto it.

In Greek mythology, **Troilus** is a Trojan prince who is killed in the Trojan War. In Chaucer's masterpiece *Troilus and Criseyde* the story is extended: Troilus loves Criseyde, who leaves him (of her own free will or against her will), going over to the Greek side in the war.

Albius **Tibullus** (*c.* 55 BC–19 BC) had relationships with other men's wives, who were also having affairs with other men. His elegies anticipate that pure love will always ultimately prevail (though that never happened in his own life).

In "They Flee From Me" poet Sir Thomas **Wyatt** (1503–1542) laments that mistresses who in the past put themselves in danger to be with him, now have no desire for contact. It's unfathomable,

> *But all is torned thorough my gentilnes*
> *Into a straunge fasshion of forsaking.*

Sir Philip **Sidney** (1554–1586), in *Astrophil and Stella* (1582), Sonnet 108, calls his "boyling brest" a "darke Furnace" within which sorrow "melts downe his lead."

Samuel **Daniel** (1562–1619) wrote the sonnet sequence *Delia* (1592) in which the continually spurned poet vows to continuously love her, even (Sonnet 33):

> *When men shall find thy flower, thy glory pass*

Indeed:

> *My faith shall wax, when thou art in thy waning.*

Closing, he warns:

> *Thou mayst repent that thou hast scorned my tears,*
> *When winter snows upon thy golden hairs.*

Gaius Valerius **Catullus** (*c.* 84–54 BC) wrote many poems about a lover he calls Lesbia. She was married and did not consider leaving her husband.

Michael **Drayton** (1563–1631) wrote the sonnet sequence *Ideas Mirrour* (1594), about unrequited love for a woman the poet calls Idea.

William Butler **Yeats** (1865–1939) proposed marriage to Irish Nationalist Maud Gonne three times, and was turned down each time. She appears in various guises in many of his poems.

In Greek myth, **Baucis and Philemon** are an elderly couple who, though living in poverty, welcome two simple but rough-looking visitors into their home, after no one else would do so. The visitors are gods who ultimately offer to grant any wish the couple might have. Baucis and Philemon request that they might die together in the same moment. Their wish is granted: after living much longer, their bodies are transformed into intermeshed trees. Ovid's telling in *Metamorphoses* is a primary source for the story.

Sonnet 4:
They'd Have Us Think

They'd have us think we think we think in lieu
of thinking truth's a form our thoughts would glean;
they'd have us think we think we intervene,
but predetermined physics must ensue.
They'd have us think we only misconstrue,
since each idea is programmed by a gene
that fate selected in the Pliocene.
How could we think at all if this was true?
    Ideas were gods in ancient pantheism—
free forms that danced across the firmament.
The chosen fix is Calvinism—but worse.
It cannot be that life's a mechanism,
for then all thought would be inconsequent.
Our love would not exist. Nor should this verse.

Notes

Though 20th Century Physics proves that the universe acts like a machine only within the limits of coarse measurement under specific conditions, the rigid machine idea of nature put forth by Classical Physics permeates modern thinking.

The idea that everything that happens (and everything that ever will happen) is completely predictable if one knows the current location, mass, and directional velocity of all particles in the Universe is still assumed by many modern philosophers, biologists, psychologists. It's as though 20th Century Physics never happened.

But when this mechanism view is extended to thought, logic collapses: if thought is mere mechanism, then nothing our thoughts express can be held to represent truth.

The **Pliocene** Epoch occurred 5.3 to 2.6 million years ago. This period was cooler and drier than the preceding Miocene Epoch (but warmer than today). The change in climate led to many areas (including the Mediterranean Sea) becoming grasslands suitable for grazing creatures. Also, the polar ice caps formed.

Calvinism is a theology named after John Calvin (1509–1564); among its tenets is predestination, the doctrine that the fate of each human soul is deterministically willed by God outside of time.

Sonnet 5:
Spirit's Validity

Perplexed by wanderings that mystify,
Copernicus denied what's clear as day;
proud Galileo labored to convey
new truths—which Urban sought to nullify.
As Bacon shaped his method, Kepler's eye
remapped the sky; absorbing their assay,
the Alchemist cast laws all things obey—
dogma was vanquished (Love pronounced a lie).
But Heisenberg, de Bröglie, Bohr, Planck, Born,
and Schrödinger confute that science; Bell
concocted inequalities that jive.
Spirit's validity has been reborn:
Physics has proven life unmeasurable—
Newton's machine lies slain. And love's alive!

Notes

The invention and development of Classical Physics created a new kind of knowledge about the universe. No longer was it necessary to put intellectual faith into a multitudinous array of unprovable beliefs (alchemy, for example). But with the disappearance of a Nature clothed in mystery, the logical possibility of other types of spiritual reality (for example, love) was challenged.

20th Century Physics broke the machine erected by Classical Physics, restoring mystery to our concept of what nature is, by re-introducing uncertainty. In doing so, in particular in identifying the correspondence between our perception and what nature "does" at the subatomic level, modern Physics restored the scientific possibility of love and life being real, as we experience them.

Nicolas **Copernicus** (1473–1543) posited that the Earth and planets revolve around the Sun and the Earth rotates once daily (the prevailing science at the time held that the Sun and planets revolve around the Earth daily).

Galileo Galilei (1564–1642) supported and defended the Copernican theory, leading to his arrest and trial for heresy.

Pope **Urban** VIII (1568–1644) was involved in Galileo being sentenced by the Roman Inquisition to house arrest for the remainder of his life.

Francis **Bacon** (1561–1626) defined methods of inquiry into nature that became the foundation of the Scientific Method.

Johannes **Kepler** (1571–1630) was an astronomer, mathematician, and astrologer whose work on planetary motion inspired Isaac Newton's theory of gravitation.

I (playfully) call Isaac Newton (1642–1727) **the Alchemist** because he considered his work in alchemy to be his most important accomplishment; it's how he expected to be remembered. And so, I grant him that honor. Aside from being an alchemist, Newton was a mathematician and physicist who formulated the laws of gravitation and co-invented the calculus.

Werner **Heisenberg** (1901–1976) won the Nobel Prize in Physics for "the invention of quantum mechanics." His Uncertainty Principle states that combinations of certain physical parameters (such as position times momentum) can only be measured to within a certain limit of accuracy (regardless of instrument precision).

Louis **de Bröglie** (1892–1987) hypothesized that all matter is describable as a wave, a key tenet of Quantum Wave Theory.

Niels **Bohr** (1886–1962) modeled atoms as having a nucleus and electrons that orbit the nucleus at fixed energy levels. He also conceived the Principle of Complementarity, the idea that natural phenomena can be scientifically represented with equal validity using seemingly contradictory logical models (for example, matter represented as being a wave, or represented as being a particle).

Max **Planck** (1858–1947) postulated that energy exists at discrete intervals, not in a continuum that spans all real numbers. In doing so, he laid the foundations of Quantum Mechanics.

Max **Born** (1882–1970) contributed to many facets of quantum mechanics, including the idea that the absolute square of Schrödinger's wave function yields the probability of observing a particle at a specific location at a specific time.

Erwin **Schrödinger** (1887–1961) conceived huge swaths of quantum mechanics and quantum wave theory. The Schrödinger Equation represents a particle as a wave or the superposition of multiple waves. Popularly, he is famous for his "Schrödinger's Cat" thought experiment, which demonstrates a core tenet of quantum theory, namely, the uncertainty as to what the state of an unobserved object is at a given moment in time: is the cat in the closed box still alive, or is it dead due to an unobserved, inevitable but unpredictable subatomic event having already occurred? Furthermore, what is the proper method for Physics to mathematically model this situation?

John Stewart **Bell** (1928–1990) devised a means for determining whether the behavior of entangled particles can be sufficiently explained by a system that includes local hidden variables. The problem he investigated was the "action at a distance" paradox that also greatly concerned Einstein and many other modern Physicists. Recent experiments based on Bell's Inequalities have shown that the results predicted by quantum mechanics match measured results beyond any prediction that could be made by any hidden variable theory.

Sonnet 6:
Nature's Secret

Daedalus crafted labyrinths, then flight;
in a wheeled horse, the fate of Troy was sealed.
Entire ships would Archimedes wield;
with Bacon's powder, kings could better smite.
What Faraday absorbed from Franklin's kite
gave Maxwell cause to codify the field.
Then Edison's and Turing's thought congealed,
yielding a strange world founded on the byte.
Manipulating macroscopic things,
controlling magnets, electricity,
induced a smug presumption: that we know.
Beneath the data, Nature's secret sings!
Ma Kali dwells there too, amazed to see
our minds enthralled by Plato's shadow show.

Notes

The greatest change in daily life for humans in recent millennia has been wrought by ever accelerating technology. We discover new things about how nature works, then invent new devices to take advantage of that, suiting human objectives.

That this all "works" on the material plane of being creates a type of smugness: the idea that we know nature simply because we can take advantage of aspects of its mechanical reality.

But Nature, in its true depth, is far beyond technology; it lives forever beyond any possible human notions. Those who have glimpsed Nature's essence, or who live within it, are surprised by our brazen modern attitude.

In Greek mythology, **Daedalus** was a brilliant architect and craftsman, who created a labyrinth for King Minos to imprison the minotaur, and wings with which he and his son Icarus sought to flee their homeland Crete.

In Homer's *Iliad*, the Greeks invade **Troy**. The Greek Odysseus designs a giant wooden horse. Greek soldiers hide inside the horse, but the Trojans believe the Greeks have left and bring the horse inside the city gates. At night, the Greeks exit the horse, open the city gates to the rest of the Greeks, and Troy is taken and pillaged.

Archimedes of Syracuse (c. 287–212 BC) was a

mathematician, physicist, and inventor. A probably mythical story is that Archimedes invented a contraption made of pulleys and other simple machines to lift a large ship entirely out of the ocean all by himself.

Roger **Bacon** (c. 1220–1292) was a Franciscan philosopher and student of empirical science. His books include the formula for gunpowder, the first such recording in Europe.

Born poor, and self-educated, Michael **Faraday** (1791–1867) became one of the most important scientists of his era. He studied electricity and magnetism, discovered benzene, and invented early forms of the Bunsen burner and the electric motor.

Benjamin **Franklin** (1706–1790) was a printer, scientist, inventor, and a founding father of the United States. His experiment with a kite in a thunderstorm showed that lightning is a form of electricity.

James Clerk **Maxwell** (1831–1879) invented the mathematical equations that form Classical electromagnetic theory. His work was based on earlier studies of electric and magnetic fields, along with his own discoveries such as the concept of electromagnetic radiation.

Thomas **Edison** (1847–1931) applied theoretical science to create new devices including efficient electric power generators, light bulbs, sound recording, the phonograph, and motion pictures.

Alan **Turing** (1912–1954) was a mathematician, logician, cryptanalyst, and theoretical biologist. He made major contributions to the invention of computer programming and artificial intelligence.

Ma Kali is the ancient Hindu Divine Mother, goddess of time, creation, change, destruction, and death. She oversees all of these as they relate to human affairs and the individual ego.

Plato (c. 428–347 BC) was a student of Socrates and Aristotle's teacher. His Allegory of the Cave describes people chained with their backs to a wall, whose only visual experience is shadows cast upon a wall in front of them by objects that pass between a bright fire far behind them, and the wall to which they are chained. Their belief, based on what they perceive, is that the shadows they see on the front wall are the actual universe, all that truly exists. The Allegory of the Cave accurately portrays the relationship between true reality (living Nature in its interior depth) and human perception.

Sonnet 7:
Hidden in the Quanta

Blackbody wavelength luminosity
defies what Maxwell, Boltzmann, Gibbs predict;
so Planck devised numerics that constrict
those hues shown wanting in intensity.
Is light a wave, or not? Planck's frequency
times h yields clumped light. Shouldn't this contradict
diffraction? Hertz led Einstein to depict
a universe of quantized energy!
Our science measures nature as it seems,
unfolding layers of phenomena,
then schematizing with intelligence.
But, hidden in the quanta, Nature teems,
not bound by any human formula,
begetting life with smiling reticence.

Notes

The path by which the inconsistencies and contradictions of Classical Physics gave way to an entirely new understanding of the universe is remarkable.

But there is, and always will be, a gap between what humans can measure (no matter how advanced future instrumentation becomes), or even conceive, and Nature's inner reality.

In modern particle physics and quantum mechanics, the measuring apparatus itself can become physically intertwined with, and even influence, the phenomenon being measured. Thus, we approach the immeasurable aspect of nature: if the measurement apparatus combined with our observation affects the measured result, then we're in effect partly measuring ourselves, our own reflection in Nature's mirror.

Nature cloaks itself in a kind of mist when we attempt to penetrate its full reality using modern logic's rigid gaze. What's on the other side we cannot know—except to know that it's somehow intimately related to our own being, our own reality.

blackbody wavelength luminosity is the intensity pattern of the spectrum of light emitted by a warm black body (for example, a wood stove).

James Clerk **Maxwell** Maxwell (1831–1879) conceived the mathematical equations that form Classical electromagnetic theory. His work was based on earlier studies of electric and magnetic fields, along with his own discoveries such as the concept of electromagnetic radiation.

Ludwig **Boltzmann** (1844–1906) developed statistical mechanics and applied this to analyze and explain the Laws of Thermodynamics.

Josiah **Gibbs** (1839–1903) worked with Maxwell and Boltzmann to invent statistical mechanics.

The theories of Maxwell, Boltzmann, and Gibbs form the ground of Classical electromagnetic Physics and Thermodynamics. In their theory, the amount of radiation emitted by a black body (for example, a fully enclosed heated oven) should increase with increasing frequency (decreasing wavelength). This theory produced a paradox where the amount of radiative energy produced at infinitesimally small wavelengths should approach infinity. Measurements of radiation from black bodies showed something different: the radiation level reaches a peak at a certain frequency/wavelength, then declines at higher frequencies (for example, ultraviolet light).

The Classical theory worked well enough to support the invention of many new technologies, such as electric trolley cars and the telegraph. But, the theory clearly did not match the observed characteristics of high frequency radiation emitted by black bodies. Physicists wondered why this was so, what was wrong with the classical theory?

Max **Planck** (1858–1947) invented the idea of energy existing at discrete numeric intervals rather than as a continuum across all possible real values, thus providing an explanation for the observed blackbody radiation spectral behavior. He speculated (based on existing experimental data) that if the energy of light is the product of a constant times frequency expressed as an integer, then the blackbody radiation high frequency anomaly would be resolved. That is, he invented a nonintuitive mathematics that "saved" (see Owen Barfield, *Saving the Appearances*) the observed data—without having any idea why such a mathematics might represent the actuality of the physical universe.

In Planck's logic, the reason there is less intensity of electromagnetic radiation at high frequencies (rather than more, as Classical theory predicts) is that a black body at a specific temperature does not create enough energy to emit as much higher frequency radiation, because more energy is required for high-frequency light to be created. That is, blue light, which is higher frequency than red light, cannot come to exist if the blackbody's temperature is too low. This explained the observed blackbody radiation spectral characteristics; but for the rest of his life Planck found the bizarreness of the idea of stepped (rather than continuous) energy disturbing.

h is the unit of conversion between the frequency of an electromechanical wave and its energy in Planck's equation $E = hv$ (where v is the frequency of the wave).

Heinrich **Hertz** (1857–1894) devised experiments that proved the validity of Maxwell's theoretical equations for electromagnetic radiation. He also discovered the photoelectric effect, wherein high-frequency light of certain colors beamed onto a shiny metallic surface dislodges electrons from the metal's surface.

Albert **Einstein** (1879–1955) was a highly prolific theoretical Physicist. Among his accomplishments was to solve the problem of why only certain colors of light dislodge electrons in the photoelectric effect. If red light is shined onto the shiny metallic surface, no electrons are dislodged, no matter how bright the red light is. But if blue light is shined onto the surface, electrons are dislodged, no matter how dim the blue light is.

Einstein's solution involved extrapolating Planck's blackbody radiation analysis regarding thermodynamic emissions of light to the photoelectric effect. He posited that red and blue light exist as a collection of individual objects, quanta, wherein each quantum has an energy dependent on the frequency of its color, as defined by Planck's

$E = h\nu$ equation.

He then posited that there is a minimum limit of energy required to dislodge an electron from the shiny metallic surface. Since blue light has a higher frequency than red light, blue quanta have more energy than red quanta. The amount of energy required to dislodge an electron from the metallic surface is higher than the amount of energy contained in a red quantum, so no electrons are dislodged no matter how bright a red light is shined onto the metal. But if even a very few blue quanta (very dim blue light) are shined onto the metal, electrons are dislodged, because blue quanta have more energy than is required to dislodge an electron. If very bright blue light is shined onto the metal, a great many electrons are dislodged.

This discovery obliterated the Classical Physics view of light being exclusively a wave phenomenon; instead, light now had to also be considered as consisting of sets of individual objects, quanta (which came to be known as photons), each quantum having a unique energy value dependent on its wave frequency (i.e., the color of the light).

Sonnet 8:
Symbols Infused

Old man, your claims are laughable to say
the least! Your "living" Nature (better than
our Science), Plato, Merlin, Thisbe, Pan—
old man, you're lost in superstition's sway!
It's really sad (almost!) to hear you bray,
old man: you promulgate utopian
visions. Who cares? You're out of touch, old man,
you blind yourself—just focus on today!
My modern son: when you communicate,
the very words you speak are history—
symbols infused with ancient life and thought.
You shun yourself, child, when you castigate
the past—the book of human memory,
a font of sight for those who would be taught.

Notes

To understand the present, today's human consciousness and our conceptions about nature and life, we must understand the past, what came before in human consciousness (which at every moment in history presumptuously considers itself to know what nature is and what consciousness itself is). We need to understand how consciousness has evolved across time, and at least have an awareness that some of the preconceptions we take as givens may be erroneous.

Our present consciousness is permeated with remnants and echoes of past consciousness, especially in our language. The vast majority of the words we use today have a very long history, in many cases millennia (most words in Western languages can be traced back to Latin and Greek roots).

Words can be conceived as being little containers of spirit that have evolved across the ages. True speech is spirit revealed. Lies are spirit (and life) destroyed.

In this poem, a young man finds an old man's depiction of the world and his trust and reverence for images from the past silly and limiting.

The old man believes his own world conception is much larger than the young man's, since it integrates the past with the present.

An historical aside (and prediction): much that is considered obvious fact today will be considered to have been silly superstition centuries from now. The future surely will look back at us in wonder: "How could they possibly have believed such stupid things?"

Plato (c. 428–347 BC) was a student of Socrates and Aristotle's teacher. Aristotle considered Plato a poet, rather than a philosopher as we conceive him today. See the notes for Sonnet 6 "Nature's Secret" for a description of Plato's Allegory of the Cave.

Merlin is a prophet, wizard, bard in legends of King Arthur.

In Greek myth, **Thisbe** and Pyramus were young lovers who both tragically died by suicide caused by a misinterpretation of events. See the notes for Sonnet 3 "Stately Elders" for a more detailed summary of the story.

Pan is the ancient Greek god of wild nature, including sheep flocks, flutes, folk music, the woods and streams.

Sonnet 9: Prodigious Dreams

Let not these Autumn notes lead you amiss.
Autumn, indeed, is come (the signs are all
around). This aging empire wanes, its fall
accelerates: decay effects abyss.
Our Winter now approaches, precipice
we didn't foresee. Our legacy? A squall
that confiscates our children's wherewithal!
Still, as we die, love, let's embrace, and kiss.
Prodigious dreams will keep our progeny
alive; though hard beset, they'll learn to cope,
and, in old age, find life is Vernal-tinted.
And generations hence will oversee
prosperity, and dwell in Summer's hope—
that fragile, fleeting dream our childhood hinted.

Notes

We look around and see the world decaying, autumn-like; then imagine what the winter will be like for our children, their children, and beyond.

While the outlook is dark, we hope that our children's dreams will keep them alive, help them endure the storm that we are partly responsible for creating; that in their own old age, perhaps they'll find new life rising out of the desolation, toward new beginning.

Some day a summer will again happen, dwelt in by some generations hence. They will see and live that summer that seemed promised but somehow died across our own lifetime.

Sonnet 10:
Cloth of Destiny

From a jail cell, he preached civility—
despite the agony centuries had brought.
Proud Rosa occupied a seat; though fraught
with risk, that act preserved her dignity.
We dreamed of equal opportunity,
and bloody segregation's end besought.
King asked: how can such justice not be wrought
in this land framed in equanimity?
  Our dreams are fractured now, disjointed; thus,
they rend King's web of mutuality—
begetting impotence. Some occupy
our parks; but do they seek to unify?
Or divide? Dr. King—jailed—cautioned us:
we weave a single cloth of destiny.

Notes

A meditation upon re-reading Martin Luther King's 1963 "Letter from a Birmingham Jail" a half century after it was written.

A sign of decay, a signal, its remarkable evidence: today there are those who consider separation and segregation of people by skin color and other obscure personal traits and categorizations to be a good thing—so contrary to King's spirit-filled vision.

Rosa Parks (1913–2005) was a black civil rights activist who defied a Montgomery, Alabama bus driver's demand that she vacate her seat for a white person because the white people section of the bus was filled. She worked with Martin Luther King toward ending segregation by race in the United States.

Sonnet 11:
Perseids

Watching the radiant Perseids above,
as lovers have through centuries previous,
embraced against night's chill—harmonious
union in nature's depth! We gaze, we love.
They pierce the sky, reflecting aspects of
mortality, flare out from Perseus
across Andromeda and Pegasus,
Auriga, Cepheus. We gaze, we love!
Teresa said our life's but hours long;
yet Bergson knew duration bequeaths clay
with pneuma, and memory's the soul's device.
We breathe our breaths; we hear the screech owl's song;
then brilliance flames across the Milky Way!
Sweet love: these hours we share are paradise.

Notes

The poem depicts the interweaving of our romantic love and nature, an eternal element in human experience and reality.

The **Perseids** is a major meteor shower which occurs annually in August.

Perseus is the constellation from which meteors in the Perseid meteor showers appear to radiate, named for the hero Perseus in Greek myth.

Andromeda, Pegasus, Auriga, and **Cepheus** are constellations near Perseus through which Perseid meteors appear to travel.

Teresa of Avila (1515–1582) was a Carmelite nun and mystic. Among her counsels is that because life is so brief, every moment should be wisely used.

Henri **Bergson** (1859–1941) was a philosopher who theorized about time, memory, intuition, and science. Among his tenets is that what science calls time is experienced in a corollary manner in our consciousness as duration; the experience of duration is a central aspect of life that cannot be measured scientifically.

Sonnet 12:
My Bliss

My bliss at dawn is touching you: in soul,
In mind, emotion, spirit, bodily.
Love, in your sleep you breathe so beauteously;
I kiss your neck and breathe—I sense your soul!

My bliss each noon is watching as you walk
This world: observing, cherishing what is.
You care for plants, and speak with chickadees!
I watch, and listen, treasuring each word.

At dusk we consecrate the day, anoint
Ourselves—this union! We are Baucis and
Philemon: gifted, blessed, we understand
This life (your goddess did not disappoint).

When midnight comes, our lives and limbs embrace—
Harmonious handfasting—our souls enlace!

Also by Kevin Farnham:

Blood and Silence (early writings), Everush Books, 2022,
ISBN 978-0-9778833-4-9.